SHOWING DAIRY CATTLE

SHOWING
DAIRY CATTLE

Bill Telfer

Farming Press

ISBN 0 85236 272 2

A catalogue record for this book is available from the British Library

Frontispiece
*Jack Rennie with the Champion Ayrshire at the
European Dairy Farming Event 1992* (Simon Tupper)

Front cover photographs
The final countdown at the 1990 Royal Welsh Show (Tim Bryce)

The British Holstein judging ring at the 1991 Royal Show (Tim Bryce)

Back cover photographs
*John Svenson-Taylor leads Grantchester Heather 8 to the Supreme Dairy
Championship at the 1991 Royal Show. Herdsman Owen Kilgannon looks on,
while Bill Telfer keeps a firm grip on the trophy* (Tim Bryce)

*Catherine Heywood leads Grosvenor Enhancer Patty, the 1992 All Britain
Champion Calf* (Sheila Metcalfe)

*Judge Edward Morgan congratulates 'Young Bill' Telfer on winning the
Championship at the Great Yorkshire Show in 1991 with Billslot Sheik
Lorraine* (Eiddwyn Morgan)

**Published by Farming Press Books
Wharfedale Road, Ipswich IP1 4LG, United Kingdom**

Distributed in North America by Diamond Farm Enterprises,
Box 537, Alexandria Bay, NY 13607, USA

Cover design by Liz Whatling
Typeset by Galleon Typesetting
Printed in Great Britain by The Bath Press, Avon

CONTENTS

Colour section between pages 42 and 43

ACKNOWLEDGEMENTS

I am indebted to the following for kind permission to use their photographs:

Alan Bishop
23 Langford Lane
Burley in Wharfedale
Ilkley LS29 7ER

Simon Tupper
15 Meadow End
Fulbrook
Burford
Oxon OX18 4BQ

Tim Bryce
Rosewarne
69 Shilton Road
Carterton
Oxford OX18 1EN

Norman Walker
Chetwynde
Liverpool Road
Sollom
Tarleton
Preston
Lancashire

Sheila Metcalfe
2 Holme Head Cottages
Dunsop Bridge
Nr Clitheroe
Lancashire BB7 3BD

Holstein-Friesian World
8036 Lake Street
Sandy Creek
New York, NY, USA
(for the North American pictures)

Many thanks to Ann Humphries for typing the initial text.

INTRODUCTION

THERE ARE many reasons for showing dairy cattle. Some people exhibit simply for the love of competition, some as a means of advertising their particular herd or herd sire, and a dwindling few enjoy it as a break from the monotony of milking twice a day.

There was a time when to win a prize or two at a top show meant that one's expenses would be covered, but those days are now long gone as exhibiting gets ever more expensive and the prize money lags further and further behind. Fortunately for the show societies there remains a hard core of enthusiasts who simply enjoy the atmosphere and the sense of camaraderie in the run-up to the judging, as many exhibitors consider the best part of the show is the time before the public are admitted.

However, the public are necessary, as without their support, there would not be any shows. Equally, and what some show societies tend to forget, the exhibitors are very important too. There is a vast difference in the attitudes of the various organisers. In some cases those running the show bend over backwards to make sure the exhibitors are well looked after, while at some places the cattle classes come a long way down the list of what is important to the show.

We can sometimes marvel at the monumental ignorance of some members of the general public, but always remember that those who know little are there to be educated. They are our customers, and while some of their actions and questions may irk a little, the public relations aspect must never be forgotten and we should never lose the opportunity to educate. Farming and its methods come under increasing scrutiny and are often subject to ill-informed criticism, and if by our answers and

Rest after labour. Maude rests after her class, while youthful admirers pose for their moment of glory

attitude we can exert influence for the good, so much the better.

Undoubtedly one of the biggest incentives for improvement is competition, and only by comparison with the best can we see what progress we are making, so take your stock and find out. Once you've decided to show, the first question is where to begin. There are a tremendous number of shows in the United Kingdom ranging from the multi-breed Royal Show to the single-breed shows or specialist shows such as the European Dairy Farming Event or the National Holstein Show. Top-calibre shows involve being away from home for a considerable length of time—up to six or seven days—so obviously only the very best of animals will make their appearance there. A little further down the scale are the county shows, often lasting three or four days, while many towns and villages have their own annual show, usually only one day.

Most journeys now can be accomplished in a few hours, thanks to a network of motorways and fast and reliable lorries. I know hauling stock—to say nothing of the fodder and the necessary gear—is an expensive exercise, but think how lucky we are, compared with the early pioneers of the show ring.

The first Royal Show was held at Oxford in 1839, and Thomas Bates, from North Yorkshire, one of the most famous of the early Dairy Shorthorn breeders, decided to take four entries. The first leg of their journey was a walk to Hull, then onto a steamship around the coast to London. This was followed by a trip by barge up the Aylesbury branch of the Grand Union Canal, and then the journey was finished on foot.

His famous bull, 'The Duke of Northumberland', slipped badly on the gang-plank when disembarking at London, but fortunately was rescued uninjured. I expect his journey was worthwhile from an advertising point of view, as each of his animals won a first prize. The whole round trip took 26 days and his bull, which weighed 180 stone when he left, was only 152 stone on his return, a loss of 3.5 hundredweight.

So here we have it. The show is the exhibitor's shop-window. Never forget that the exhibitor is judged as well as the exhibit. When the judge has made his final line-up, remember that that placing is simply one person's opinion on the day. My maxim when judging has always been, 'I may not be right, but today I'm official.' If you should happen to win, be modest in victory. If you lose, don't kick up a fuss—nobody loves a sore loser. Most judges are people of integrity and will place an animal on its merits without consideration for the person on the halter.

To those of you thinking of exhibiting for the first time, don't

A nice line-up at the English Royal Show (Tim Bryce)

be too disappointed if your animal stands down the line. Many first-time exhibitors discover that what they think are swans at home are only geese when they get to the show. Most judges worth their salt will walk down the cattle lines after judging. They'll be happy to give advice. Stay by your animals after judging, and don't be afraid to discuss them. Many a sale has been made to someone who perhaps liked your animal a little better than the judge. To the really committed, the challenge from that time forward is to breed something better, as hopefully you will have enjoyed the friendship at the show and you will want to go back.

I hope you don't find too many mistakes in my advice. We made many at first, but thanks to the good friends gained at the early shows we attended, and the good advice freely given, we slowly moved from the bottom of the line to the top. To all those who gave that advice, and to the many other friends we have made at the shows over the years, I am deeply indebted for most of the material in this book.

IDEAL CHARACTERISTICS

ALTHOUGH THERE are several different breeds of dairy cattle, the ideal characteristics are virtually the same for all breeds, with variations as to size and colour. There is still truth in the old adage, 'If you can successfully judge one breed of dairy cow, you can judge the lot.'

Our ideal show animal should not only be a good type but she must also be profitable. We must never forget that the first essential in a dairy cow is milk, and the ability to produce lots of it profitably. The cow we are looking for nowadays is one which justifies its existence from an economic standpoint by returning the greatest possible profit. Raising heifers is an expensive business and the longer a cow can profitably stay in the herd, the more money she will make. She must have the inherent ability to produce heavily.

Without this ability it is most unlikely that she will ever figure successfully on the show circuit, as correct conformation and strong constitution combine to make a desirable type. It is most probable that any animal which does not milk well will be lacking in dairy character and bone quality, characteristics which are very important in the show ring. It is by now a well-established fact that animals which score highly on type tend to last much longer. Let us therefore look at the traits which make up our ideal animal.

When I was young, I always reckoned to look at a young lady from the legs upwards, but with our dairy cow I guess we must start at the other end. The general appearance should be attractive indicating femininity, vigour, strength, size and stature, the whole lot combining harmoniously and appearing to be in proportion. On the move the carriage should be alert, stylish

Everyone's favourite: Cardsland Matt Delila, Champion at the Royal Show 1989, and one of the best black and white cows bred in Wales

and impressive.

Dairy character is a combination of many factors which together give that extra dimension. The head should be clean-cut, with the eyes large and bright. The ears should be carried alertly. It is also desirable that the head should be long, as there seems to be a definite correlation between the length of the head and the ability to milk. An animal with a short head never looks completely correct even if the rest of her component parts are. The neck should be long and lean and nicely blended into the shoulder. The throat should be clean and the brisket should not be too heavy. Cows with a heavy brisket are seldom good producers.

The withers should be well defined and wedge shaped while

the ribs should be wide apart with the bones wide, flat and long. Flanks should be deep and refined while the thighs should be flat from the side view, but wide apart when viewed from the rear, with plenty of room for the udder and its rear attachment. The skin should be thin, loose and pliable and covered with fine hair. The udder must be soft and pliable and free from meatiness, while the bone should be flat and clean.

The things which combine to give dairy character must be combined with capacity, which gives the animal its ability to produce. The muzzle should be broad, with open nostrils. The jaws should meet properly, as parrot-jawed animals have difficulty grazing properly, and this will be discriminated against by a competent judge. The shoulder blades should be set smoothly against the chest wall and withers, blending nicely into the body.

The chest should have a wide floor, giving ample width between the fore legs. There should be plenty of girth for heart room. The heart is the motor which drives the whole machine so it is important that it is not cramped. The animal should be well filled at the crops. The back should be long and straight, with the vertebrae clearly defined. The loin should be broad and slightly arched with a high and wide attachment at the hip bones.

The feet and legs are very important as only when they are correct is it possible to get proper carriage on the animal. Ideally the feet should be short and well rounded with a deep heel. The toes should be slightly spaced. The pasterns should be strong and flexible. The fore legs should be straight and wide apart, with the feet placed squarely. There is a tendency for the front feet on some animals to be severely turned. This is a major fault.

Hind legs ought to be nearly perpendicular from hook to pastern from the side view, straight and wide apart from the rear view. Hocks should be free from puffiness. Legs which are badly sickled are a major fault, as are legs which are too straight and posty. Other faults in the legs are crampiness and signs of arthritis. The rump should be long, wide and clean, blending nicely with the loin. Hips should be wide but not prominent with the pin bones wide apart and free from patchiness. The tail head

A very weak loin

should be refined, level through with the back bone and set slightly higher than the pins. The tail should be long and slender. An advance tail setting and a wry tail are discriminated against.

The udder should be strongly attached and well balanced, with a fine texture. It should be symmetrical, of moderate length, width and depth. There will be slight quartering on the sides. The median suspensory ligament should be strong, showing a definite cleavage between the halves of the udder. The udder should be soft, pliable and elastic and should collapse after milking. The fore udder should be firm and smoothly attached to the body wall, of moderate length and with the quarters evenly balanced.

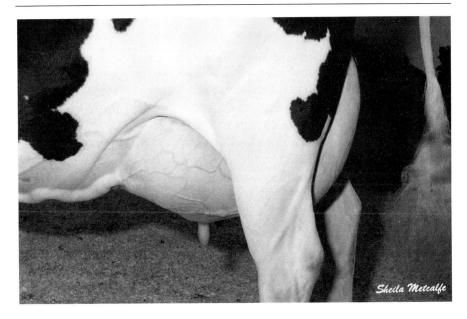

They don't come much better. The business end of Grantchester
Heather 8, Best Udder National Holstein Show 1992 (Sheila Metcalfe)

The rear udder should be high, wide and strongly attached. It ought to be slightly rounded, of uniform width from top to floor and with the quarters evenly balanced. Teats should be medium length and of uniform size. They should be cylindrical and plumb. Viewed from the side, the teats should be placed in the centre of each quarter. The rear view should show the teats slightly closer to the inside than the outside of the quarter. Milk veins should be long, tortuous and well defined while udder veining is desirable.

Faults in the udder carry different degrees of penalty. For example, a blind quarter will mean disqualification of the animal, as will abnormal milk. Other faults attracting penalties from slight to serious discrimination are side leaks, injured teats, light quarter, hard spots in the udder, webbed teat and obstruction in the teat.

Other more general things which carry discrimination are lack of size for the breed, lack of stature and over-conditioning. Temporary minor blemishes will also be noted. Freemartin heifers are not eligible for showing, and evidence of sharp practice will usually result in disqualification.

HOUSING CALVES

FOR THOSE of you well versed in rearing calves the next sections may be an irrelevance. However, housing and feeding play important parts in the development of your budding show champion, as faults in these systems can cause illness and disease from which the animal may never quite recover.

Clean buildings are an essential. I know it isn't possible to start every calf off in a building which hasn't held calves before, so make sure your calf building is thoroughly cleaned and disinfected on a regular basis. Many farms now boast purpose-built rearing houses with automatically controlled temperatures and ventilation, but these are generally outside the resources of smaller farms.

Don't despair if you don't run to one of these calf palaces, as by using a little commonsense and ingenuity it is possible to get very satisfactory results without one. The main requirements of your calf are that it should be warm and draught-free, and these basics can be achieved without too many fancy materials. If you have no individual pens, it is quite easy to create very acceptable conditions by simply using straw bales for the sides of the pens.

If the weather is very cold a roof of bales can quite easily be made. The big plus to this system is that when your calf is big enough to move to the next stage, the bales can be used for bedding adult cattle and new pens constructed for the next bunch of calves. This greatly helps to reduce the incidence of disease. In low buildings look out for condensation as this is usually a sign that ventilation is inadequate. Plenty of straw on the floor will ensure that your calf is nice and cosy, but your straw must be dry. Damp, soggy straw is a certain way to mischief.

11

It is possible to have slats underneath the pens. This has the advantage of allowing the urine to drain away, and it uses straw economically. On the other side of the coin, too little straw on top of the slats can lead to leg troubles, and it certainly leaves much to be desired from a comfort point of view. Plenty of dry bedding is a must all through the life of your potential champion. This ensures that your animal keeps reasonably clean. An animal which spends her life in a system which is chronically short of bedding will inevitably become stained, and it is virtually impossible to completely remove the stains once this state has been reached.

Most show calves will spend much of their lives in individual pens, even when they get older, so make sure the pens are always well bedded. Most pens nowadays have concrete floors and if the pen is large it is a good idea to make a bedded area at the back and to clean out the rest daily. This system makes for good feet, and the concrete is much better for them than continually walking on a soft bed.

Youngstock in pens will get enough exercise, but remember that any picked out for showing need regular handling, not so much for the exercise as for the education.

There is now a large range of purpose-built hutches for use outside. These seem to work reasonably well, but I have no personal experience with them. They are widely advertised, so it should be possible to inspect some in use if you are interested.

FEEDING CALVES

NO MATTER what the age of the animal, feeding plays a very important part in the preparation for a show. It is important that stock are well reared as, whether you are showing or not, a sound rearing period will stand your animal in good stead in later life. It's no use taking a badly nourished youngster and expecting to bring her to reach show condition by a short period of excessive feeding. That is like taking an ill-nourished boy and expecting to make him fat with a couple of meals of steak and chips.

There are several good traditional ways to rear calves, and providing your system is sound and the calves you rear are healthy, that's fine, simply start that way. Do not, under any circumstances, over-feed. The object is to develop an animal which is well grown for its age. Over-feeding simply makes your calf fat, and an over-conditioned calf stands little chance under a competent judge. All fat does is to disguise the true shape of the animal, and if carried to excess it may settle in the mammary system, creating disastrous results in later life.

Although some exhibitors of beef cattle think it is a very good thing to suckle their calves, *never* be tempted to do this. A dairy heifer which has been suckled as a calf is very seldom any use in later life as the fat which seems to be laid down in the udder tends to inhibit maximum production. Calf-rearing is excellent training for any erstwhile exhibitor because the successful calf-rearer will have learned that strict routine and attention to detail are very important. So too is the underrated art of observation.

Whatever the favoured feeding regime, there is one thing which is of vital importance—colostrum. Colostrum must be fed in some form or another as soon as possible after birth for at least four days. Some farmers favour letting their newly born calves suckle their dams for two or three days, but if the calf is to be bucket fed, it's much better in the long run to remove her from her dam as soon as she is nicely licked and dry and then to ensure that she gets her proper colostrum ration via the bucket. Calves which have been suckled for even a couple of days are often reluctant to take to bucket feeding, and this causes quite a bit of stress to the calf and sometimes considerably more stress to the person trying to teach it.

The colostrum should be fed at blood heat and the first feed ought to be given within six hours of birth. If this can be carried out sooner, so much the better. The calf should have four pints (2.27 litres) per feed and should initially be fed twice daily. It goes without saying that the bucket used should be clean and regularly sterilised, as the digestive systems of baby calves are notoriously easy to upset. While it is possible to treat many disorders, some, while not killing your calf, will certainly stunt her growth and spoil her potential forever as a show animal.

After four days your calf should be gradually changed onto your favoured rearing method, either whole milk or milk sub-stitute. Both do a good job provided they are sensibly carried out. I have seen some farms where the calves are fed the milk from cows treated with antibiotics and unsuitable to go in the tank. This is very bad practice and can cause all kinds of problems. If using milk powder, read the instructions and stick to them. More calves have been killed by over-feeding than by under-feeding.

Calves should be encouraged to take solids as soon as pos-sible, and the bucket with a proprietary coarse mix or baby-calf pellet can be hung on the pen front next to the milk bucket. Roughage is also important from an early age. If your calf is bedded on clean straw, this is usually enough to start them off, but make sure they have a little clean straw daily. These early days are very important in the development of the rumen, which is crucial in later life in consuming and converting large quan-tities of roughage.

There was a time when soft hay was reckoned to be an essential for young calves, but the vast majority of calves reared now never see hay. After about three weeks offer a little cold water daily, so that the calf is well used to it before she is weaned, as some calves will drink to excess when first given the chance. By the time your calf is six weeks old she ought to be eating enough pellets or mix for the milk to be discontinued. For the last week before weaning, substitute water for half the milk and then the change-over should be trouble free.

Once the calf has been weaned, she will start to eat a little silage daily. The pellets should be continued until the animal is about six months old if she is staying inside. Around four pounds (1.8 kg) of pellets daily will keep the animal growing nicely provided silage is being fed ad lib. If you have your show candidates picked out, a pound or two (.4–.8 kg) of linseed cake daily is very good for putting extra shine on the skin.

There are almost as many ways of feeding cows as there are cowmen and the same applies to feeding youngstock. All, however, are geared to the same goal, namely to keep the young animal growing without getting fat. With the trend nowadays to calving heifers at 24 months or even younger, it is imperative that your youngster does not suffer any setbacks, as a bout of disease or a month on a poor diet can very well mean she will never really catch up with her contemporaries.

The aim then should be for consistent growth without any setbacks. In Britain almost all animals will spend at least one summer at grass and many of them will also spend at least part of a second. For instance, a calf born in October would be inside until early May and then would be at grass until late August when she would be housed for the winter. She would then go back to grass in mid-April until she was housed again in September, ready for calving down the following month.

We have a rule of thumb that anything born after January 1st will not go to grass in its first season, only going out in April the following year. This seems to work quite well, as calves which are put to grass too early are very susceptible to stomach worms and husk, which are both potential killers if neglected. All animals at grass should be wormed regularly—it's surprising how quickly they can get reinfected. For a few years now we have successfully used a proprietary slow-release bolus prior to turning out. This will give protection for the entire grazing season. The cost of these boluses is fairly tooth-numbing, but they are probably no more expensive in the end when one considers the accumulative cost of the routine dosing and the sheer physical cost of bringing the animals in on a regular basis. In North America many of the youngstock never get out to grass, but are reared indoors.

Supplementary feeding while at grass should not be necessary. Bring animals intended for the show team inside a few days ahead of the first show so that you can get them onto their show diet. This will probably be hay, coarse mix and sugar beet pulp if you live in Britain. In North America maize silage, lucerne and cotton seed will form the bulk of the diet. Linseed cake should be fed daily throughout the show season.

Ad lib silage will keep animals being brought on inside going nicely once they get to about six months old. Any to compete in the show team will benefit from a little linseed cake as detailed above.

MANAGING ADULTS

B Y THE time your show animal becomes adult, she will probably have entered the production herd. How she is treated then is very much a matter of how much is available and what is preferred by the owner. In summer most dairy herds in Britain live outside at grass. This is a very convenient way with the cows getting plenty of exercise and having a soft bed. Troubles such as swollen hock are virtually unknown. In North America, however, the system is entirely different, with most cattle being under some restraint, with only exercise paddocks, and food being offered inside.

In Britain in winter most dairy herds are maintained under a cubicle system—that's free stalls to those of you who live in Canada or the United States. A proportion are wintered in straw-bedded yards, while a few are still in byres or tie-barns. Some owners simply leave their show animals with the rest of their herd mates between shows. This system has the great virtue of being simple and makes for uncomplicated herd management. Provided the animals are happy and 'fit the system', this way of doing things has much to recommend it.

Many farms, in fact, would find it almost impossible, from both a housing and a labour point of view, to keep show animals on a separate regime. There are some farms where circumstances allow the show team to be managed as a separate entity, a Utopian system which I personally never managed to implement.

The system the animals are kept on dictates the events pre-show, as the bowel of an animal which has spent several weeks at grass is hardly likely to win too many friends at a show. Any grass-fed animal must have at least three or four

17

days on its show diet prior to leaving home. This will ensure that there will be less stress on the animal's stomach. It ought also to mean that it is possible to use a shovel to catch the ordure rather than the bucket which would be a necessity for the grass-fed animal.

Obviously any animal which is out at grass or on a free stall or straw yard system will be getting enough exercise without the need for special exercise sessions. However, it does no harm to slip the halter on occasionally as a reminder of the facts of life.

Animals which spend most of their lives in byres will need a lot of exercise and if possible an exercise period should be aimed for every day. Twenty minutes' walk daily will keep your candidates' muscles in good order and banish the spectre of puffy hocks. People who are able to keep their show team in box stalls usually have plenty of labour to make sure they are given every attention. If you haven't got the facility of box stalls or the extra staff needed to run the system, don't despair, as many of the recent champion animals in Britain have, apart from show times, been treated simply as 'one of the herd'.

TRAINING CALVES

M Y GRANDFATHER used to say, 'Start them young.' That advice is still relevant today. Not only that, but it applies to both the handler and the calf. One of the big differences he would notice should he come back would be the number of classes for calves as well as for handlers. One of the major steps forward in standards of exhibiting has come about from the introduction of calf shows.

This is something we in the United Kingdom have copied from North America and it gives splendid grounding to both the exhibitor and the calf. The following 'rules' apply equally to adults.

Dairy cattle showmanship is defined as 'the art of displaying an animal to its best advantage'. The animal always looks better if it is properly handled and controlled, so we'll deal with the calf first.

There is much more to showmanship than at first meets the eye as behind every good showman lie hours of work and preparation. At any calf show it is very obvious which of the contenders have put in lots of background work. Calves do not lead nicely and show themselves properly unless they have been well trained. A calf which continually fights the halter never shows herself to her best advantage and she certainly does nothing for the showman's chances. Many a potential prize-winner has ended up well down the line because the judge has been unable to assess her properly.

When selecting a calf to work with, look first for one with style. This is a quality which is very difficult to describe, as it is the overall impression of the calf rather than one particular component. The stylish calf will show plenty of dairy character;

she should have a good, strong level topline and have good feet and legs with nice flat bone. The head will be carried high and the whole outlook should be alert. Also, the calf needs to be well grown for her age, as very small calves stand little chance in strong competitive classes.

You'll find that calves are very like people, and there is a wide variation in temperament. Hopefully the calf you have picked to work with is blessed with a good temperament, as occasionally one finds a really mean animal that is uncooperative and bad-tempered. If patience and kindness do not bring about an improvement, it is probably better to get another calf, as, while it may be just about possible to handle an obstreperous animal when she is small, it becomes well nigh impossible when the animal is mature.

Some animals are so unstable that it is never possible to get them properly under control, and nothing makes an exhibitor more unpopular than being responsible for a mis-stitched animal charging around the show ring upsetting all the others. I'm sure we've all occasionally seen such an animal, whites of eyes showing, tails switching madly, dragging the unfortunate leadsman around the ring. Just make sure it isn't one of yours.

There is a great advantage in breaking all calves when they are very young if time can be found. A calf will easily come to terms with being fitted with a rope halter and tied up for a few minutes at a time. Tie the calf with her head up at about the height at which she will be led, and keep watch initially to make sure she doesn't hang herself.

After she has had a few days tied up, start leading her. At this stage make your selection. If your calf is young you should have no difficulty in teaching her to walk with her head high and to parade gracefully. She should very soon respond to the halter without undue exertion from the leader.

If the calf has grown very strong before she is haltered, some people use a tractor in the early stages of training. The advantage here is that the tractor is much stronger than the calf, so there is no danger of the calf pulling away and escaping, as once an animal has discovered that she can get free, this becomes a habit which is very difficult to break. Fasten the halter onto the fore end of the loader, and raise the loader to the height which will bring the animal's head to the height at which it will normally be carried while walking in the show ring. Now move your tractor slowly in reverse so that the calf is forced to follow. If she is smart she will soon discover that it is easier to walk than to hang back. If you have a calf that leads properly, it is a good idea to tie her alongside your novice calf so as to calm her.

While you are doing this, make sure someone is with the calf at all times. The reason for this is two-fold. Firstly, if the halter is held loosely when the calf is pulling, the animal will get the notion that it is the handler rather than the tractor that is making it impossible to pull away. Secondly, if the halter becomes too tight under the jaw or the calf throws herself down and chokes, someone is there to loosen the halter.

Keep the sessions fairly short at first, twenty minutes to half an hour maximum, and then gradually lengthen them. This prevents the calf from becoming bored. I always like to give a small feed of coarse mix or something similar as a reward at the end of the session. This makes the calf associate the halter with pleasure rather than unpleasantness.

Whatever you do, resist the temptation to treat your calf as a pet. It is vitally important that she knows that you are the boss. Be kind and gentle with her of course, but do not pet her. Petted calves have a very nasty habit of making fools of their handlers when they get into the ring.

After the calf is moving nicely, it is time to introduce her to the leather halter, which is a must in the show ring. Make sure you use a proper calf halter and not a cow halter, as the nose-bands on these are far too large. Never, *never* use your leather halter to tie your calf up; that's not what they are made for. Also make sure you train your calf to the leather halter well before the show. This feels much different from the rope halter, especially the chain under the chin, and some animals take exception to them in the early stages.

As well as teaching your calf to lead, it is also important that you teach her when to stop, as signals from the judge need to be acted on quickly. It doesn't gain too many points if, in spite of your efforts, the animal continues for several yards after the stop signal. There is even a right way and a wrong way to stop, and if you can teach the right way when the calf is

21

young, it saves much time and bother later.

Ideally, when you stop, the front legs should be nearly in line, with the one nearest you slightly behind the other. The back leg nearest you should be slightly forward of the other. The head should be held high and tilted slightly towards the judge. If the front feet are not quite correctly placed it is acceptable to gently tap or step on them with your foot, but the back legs should never be moved in this way.

If the legs are incorrectly placed, some pressure on the halter, either forward or backward, will usually have the effect of moving the feet into the required position. If some pressure is applied to the point of the shoulder with the fingers of the right hand at the same time as the pressure on the halter, this will help to move the calf backwards.

By the time you and your calf have progressed to the show ring, you should be aware of any faults, and when you 'set up' the calf you should do it with a view to correcting these faults as far as possible. If the calf is 'throaty', i.e. a bit heavy through the throat, it helps to take some of the skin in the right hand and draw it gently upwards. This makes the animal look much cleaner right through the head and neck region. Similarly, a weak loin can be improved by positioning the back feet well underneath the body, while a high topline can be improved by touching down just in front of the hook bones.

PREPARING CALVES

IN THE ring your calf should look her very best, with every-thing that can be done to this end having been dealt with properly. It is important to start well in advance, as it is impossible to get an animal to its best condition in just a few days. Good husbandry should have made sure that there are no lice; if there are, get them killed in good time as they rapidly remove the bloom from an animal. Check that there is no ringworm, and if there is get it treated, as no animal can be shown which has signs of active ringworm.

Should the feet need a slight trim—as many do when they have spent a long time in straw-bedded pens—be sure to trim about four weeks before a show so as to allow plenty of time to get over the tenderness which inevitably ensues.

The animal should be washed several times with soap and warm water, or with one of the many shampoos made especially for the job. The idea of this is to get the hair into good condition ready for the clipping. It is very important to make sure that the soap or shampoo is properly rinsed away so that no traces remain, as anything left tends to leave the hair looking lifeless instead of nice and shiny, which is the object of the exercise.

A power hose is a very useful tool for rinsing the coat but be sure to reduce the pressure a little before you use it. Too much pressure can cause pain and discomfort to your calf. Be especially careful when washing the head and ears and it is better not to use a hose here. Soap in the eyes is very painful and try not to get water in the ears, as this can cause them to hang down for days afterwards. A soft rag is very useful for wiping the inside of the ears.

The animals will have been clipped six weeks before the show, but many people clip their animals again a week or so ahead of the show, with the final trimming made at the showground itself, as most of the major showgrounds now have electric points laid on. The days when everyone clipped their own stock are long gone and several professional fitters travel the country. These persons—I don't say men, as there are also one or two very good ladies—will travel to the farm prior to the show, or will do the clip job at the show after the cattle have arrived and settled down.

Clipping an animal well requires lots of practice. If you are inexperienced, it is a good idea to get a professional initially so that you can get a notion of what is required. After that it is simply a matter of practice, although, as in most things, some people seem much more adept at it than others. If you can, practise first on calves which you don't intend to show; then it isn't too important if you make a bit of a mess. As with most other things, practice makes perfect.

SHOWING CALVES

N EXT WE come to the leadsman. First of all, the leadsman should be neat and clean, dressed according to whatever the relevant breed society recommends (see page 41 for full details). Smoking is a definite no-no while in the ring, and many judges do not like to see competitors chewing gum.

The way you dress very often reflects the way your calf is turned out, as a sloppy dresser will very often present a calf which is less than satisfactorily prepared. Keeping clean in the last minutes of preparation before entering the ring can be a bit of a problem, so if you can, wear a boiler suit on top of your show garb until the last moment.

The showman should at all times be alert, yet polite and courteous to the judge and fellow competitors. It is fatal to lose your temper, so emotions must be kept under control. The showman should be keen yet relaxed and observant. Keep the animal under control, but be aware at all times as to where the judge is, and when he signals react to his commands immediately. Be prepared to answer any question the judge might ask. It is important when showing calves to know the date of birth, and the leadsman ought to be able to tell the judge when a cow last freshened and how many calvings she has had.

Get to the ring on time! Some shows insist animals parade in age order, and some in catalogue order. These only involve getting into line when told by the ring steward. Some, however, have no set order of entry, and if you find a show like that, try to get into the ring first and create an early good impression. If you are first into the ring always travel in a clockwise direction unless you are instructed otherwise, and whatever you do, don't chat to the other showmen or the spectators around the

Callum Mckinven casts an eye over a calf class, 1991 (Tim Bryce)

ringside. You have enough to do watching and obeying the judge without any other distractions.

Enter the ring holding the halter in the right hand and walk in smartly. The judge wants you in as quickly as possible so that he can get started. Once well into the ring, turn around and take the shank in the left hand, with the hand close to the halter. Hold the surplus shank in two loops. By holding the shank close to the chain you will have more control, as you can slacken or tighten the chain with the minimum of effort.

As you are travelling in a clockwise direction, this enables the judge to see the whole animal without the handler obstructing his view. When all the competitors are in the ring, you should, if you have not already done so, have changed hands on the halter and be walking backwards. Try to move in time with your calf; when she stops, so should you. The best handlers seem to have the knack of always appearing to be in tune with their calf. Much of this is down to practice beforehand. Keep your eyes on the judge, but you must always be aware of what is going on around you. I often think that eyes which worked independently would be an advantage at this stage.

The judge will probably ask you to stop and start several times and he will be observing how quickly you respond to his signals and how you set your calf up. Use your right hand on the point of a shoulder for extra control. If your attention wanders and you miss a signal, this will usually mean that you drop several places in the final line-up.

After these initial manoeuvres, the judge will normally examine each animal individually to see how you perform on the halter. Never relax when the judge is examining someone else, as judges have a nasty habit of turning round to see what is happening behind their backs. When it is your turn to be examined, bring your animal forward and set her up quickly. The judge will feel the skin, and when he does this, turn your animal's head slightly towards him. This loosens the skin, and fine loose skin is a sign of dairyness.

The judge will then usually walk around the back of the animal and look along her back from tail to head. You should by now have her head straightened up again. Sometimes he will feel the skin on the other side, and if he does, remember to turn the head towards him again. He will then walk around behind you and look at the animal from the front. Step away from the animal slightly and make sure the head follows him round. When all the animals have been examined the judge will make his line-up. When you are called in, do so as quickly as possibly by changing hands on the halter and walking briskly forward.

If you are called in first, the steward will show you where to stand; if there are animals in the line-up already, keep in line. If your calf moves forward, try to make her back into line with pressure on the shoulder. If this is not possible, pull her right forward and start again, although this may possibly lose you points. In handling classes, especially if the places are close, the judge may ask three or four of you to swap calves to see how aware you are of the faults on another calf, and what efforts you make to minimise them.

If asked to change places in the line there is a correct way to do this and it should be done as quickly as possible. Walk forward until your calf is well ahead of the line, turn clockwise, come back through the space you have just vacated and then move into the new position. Give the animal plenty of room to turn and remember that the leadsman always goes round the animal and not the animal round the leadsman. The only exception to this rule is if you are moved from first to second in which case turn anti-clockwise and come immediately into your new position.

The correct movement from fourth to second place

The correct movement from first to second place

The question of attitude is very important and is too often neglected. The judge will be taking careful note of your attitude, to both your calf and your fellow competitors. Don't crowd the animal in front, and if the leadsman behind crowds you, ask him nicely to keep his distance. Sometimes feelings run high, so resist the temptation to get involved in a slanging match. If your animal should play up, it is very important that you stay calm. Nothing looks worse than an exhibitor punching an animal in an attempt to bring her to order. It solves nothing and usually upsets the animal even more. Not only that, but any judge worth his salt will undoubtedly penalise you heavily. If, in spite of all your training, your animal does misbehave, you have a much better chance of bringing order out of chaos if you are quiet and firm. If you stay cool and regain control, it shouldn't ruin your chances.

Look cheerful! It costs nothing to smile and never forget that in handling classes 20% of the marks are given for attitude. Your calf may be slightly less than perfect, but in these classes the other marks are 40% for preparation and 40% for technique. You have control over all these things and where you finish in the line-up is usually a good indication of how good your pre-show efforts have been. Don't be too disappointed if you didn't do quite as well as you had hoped as not everyone can be at the top of the class. When the judge gives his reasons at the end of the class, remember what he said about your efforts and if he criticises any particular part, make a special endeavour to correct that next time.

DOING PAPERWORK

IF YOU have never showed before you will find that there is lots of paperwork involved. For a start, having decided which show to attend, you will need to make your entries, so the first thing to do is to apply to the show secretary for a schedule of classes and an entry form. Having received these, look first for the date when entries close. *This is important.* If you fail to submit your entries by the date specified, it is very likely that they will be refused.

Next, check to see what the conditions of entry are. Some show societies insist that details of each animal should be submitted with the entry form. If the closing date for entries happens to be six or eight weeks before the actual show itself, this can create problems as it is most unlikely that all your chosen candidates will be in satisfactory condition then. If, for instance, you decide to enter three classes, it will be necessary to nominate at least two entries for each class and hope that one of them is suitable when the time comes.

This is quite a time-consuming operation, as you will be required to fill in all relevant pedigree details on the entry form, to say nothing of having to pay an entry fee for each animal, even though you know that only half of them will get there.

Some of the more enlightened societies now have a much better system. If you want to enter three classes, then you simply book three stalls and take whatever animals you think will do best, specifying which classes you wish to enter when you get there. You will, of course, be expected to furnish details of each entry at that point.

For each animal you will need an official registration certificate issued by the Breed Society. In many cases you will also need an up-to-date lactation certificate for each animal in milk. If the animal is too young to have calved, a certificate for the dam is useful. Most shows now specify that the entries shall be from a herd tested and officially free from enzootic bovine leucosis (EBL) so you will need a completed form to that effect. Also, if any of the animals have had ringworm in the recent past, a veterinary certificate to the effect that the ringworm is now dead is absolutely essential.

For the in-milk classes you will probably find that an animal which has calved within the last twelve weeks or so will be in the best show condition. Usually a two- or three- week period immediately after calving will find the cow's udder full of 'nature', but once that has gone, she ought to be in full milk and in full bloom. The seasoned exhibitor will have made plans the year before and taken steps to ensure that his animals calve at the correct time. However, with the improvement in genetics, some animals nowadays milk on and stay in condition for a much longer period than used to be the case.

I suggest that you carry all these forms in a large envelope or a folder, not in the show box, which will quite often be inaccessible until after the cattle have been unloaded. At many shows the animals are checked as they leave the truck, so you can see why a folder is more convenient.

Try to get all the papers together in good time, as there is nothing worse than looking for these at the last minute. You will have plenty of other things to occupy your mind then.

One final word of warning about paperwork: do not forget it. As it isn't in the show box, it is very easy to do so.

CLIPPING

WHEN I first showed cattle, clipping presented no problem, as no one did any more than simply clip the rough hair from the tail and tail head and possibly some from the fore udder. Mind you, the accepted tool for the task then was a pair of sheep shears, so it's not really surprising that the job was somewhat rudimentary.

The advent of electric clippers, plus the skill of some operators, has completely transformed the operation, and many people, some of them quite young, are now able to turn out cattle looking immaculate.

There are several makes of cattle clippers on the market, so if you haven't got a pair already, have a look around any show; watch the exhibitors at work, ask questions as to how good the clippers are and make your choice. Don't think for one moment that a new set of clippers will result in you doing a good job immediately, as much practice is needed. Try clipping a cow or two at home which you do not intend to show. A bad clip job takes longer to grow out than a bad haircut and the results are usually there to be seen for quite a long time.

The object of clipping is to enhance the appearance of the animal and to have her appear natural when she enters the ring. It should help display an animal's full dairy character and physical strengths so that she shows overall balance and style. A good clip job will highlight her strengths and will hopefully help to camouflage her weaknesses. You should clip your animal all over six or seven weeks ahead of the show so that it will be easier to make a good job before the show.

Before clipping begins, wash your animal thoroughly, as described earlier, and dry her. Before you begin, stand back from the animal and have a good look at her. Note her weakest points and bear these in mind at all times as you will be endeavouring to enhance her appearance. Very few animals are the same, so each may need a slightly different approach. Make sure your clippers are sharp and have the correct blades. Don't hurry. Hasten slowly, as you can always take a little more off later.

The correct technique involves cutting with the hair growth or against it, but not across. Different parts of the animal need different approaches. If you have never clipped an animal before this will seem very complicated, but hopefully you will soon master the job. Remember, we are trying to emphasise the femininity of dairy animals and everything is done with that in mind.

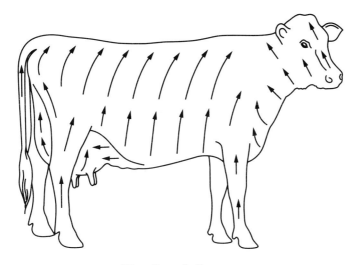

Direction of clippers

The first time any animal comes face to face with a noisy set of clippers must be a fairly big shock to the system, so don't rush in too quickly. The tail is probably the best place to begin to get the animal acquainted with the noise and feel of the clippers. Get hold of the tail and find where the switch begins; if you look first you will be able to see where the long hairs start. Do not cut the switch, as it needs to be left long, but cut upwards from the switch, working towards the tail head, blending into the top line.

Once having got the animal used to the clippers, most people prefer to start at the head and work backwards. The hair should all be carefully removed from the head, taking care to follow the natural contours. The inside of the ears should be cleaned off with a set of dog clippers, which are a useful addition to the normal clippers for close work. Removing the hair, especially around the poll, emphasises the dairy character, as a long head always looks more feminine than a short one.

Clip the neck, shoulders and brisket, blending the hair through the withers and the shoulder blades, by clipping halfway across and blending into the rest of the hair. Clean off the brisket, as a heavy brisket spoils the look of a dairy animal. Avoid clipping hair behind the shoulder.

The topline is a tricky bit, as the line of the spine is not always level, and there is no doubt that a level topline makes for an attractive silhouette. From the withers back to the tail head brush the hair up from each side so that it forms a wedge shape about three inches wide at the base. It helps to blow dry the hair in the position. Stand back again to see where the trimming needs to be carried out to produce the desired results and then gradually trim off the bumps. Some people use scissors for this while others prefer the clippers. If using clippers, keep the back of the blade to the hair, rather than the front, as this should prevent too much hair being taken off at once. On show day this wedge will need to be brushed up again and held in place with a little spray adhesive. Blend the root of the tail into the topline.

The back legs should be completely clipped out to highlight the bone quality. This clipping should be carried on and blended in at the second thigh. If the leg has a tendency to be sickled this can be camouflaged to some extent by leaving some hair on in front of the hock and trimming close at the back of it. Similarly, if the leg is too straight, take all the hair from the front and leave a little at the back. It's surprising how the shape of the leg can be altered in the hands of a skilful fitter.

The udder is often left till last but is very important. Clip out the rear udder first, taking all the hair off until you come to the sides, where it needs to be blended into the body and legs. The way the fore udder is clipped depends very much on its shape. If the fore udder is well attached, it can be fully clipped out. This

34

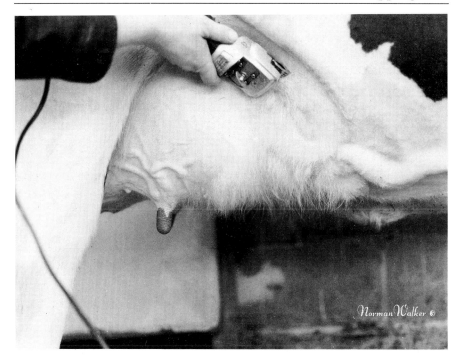

Now you see it—now you don't. The milk veins come into view as the long hair falls away (Norman Walker)

highlights the veining. If the fore udder is short, however, blend the hair on it into the rib, which gives it a longer appearance. Clip the milk vein out as far as the well and blend the hair at the sides. A good job on the udder and milk vein can make the difference of several places in the final line-up.

If the animal has been clipped six or seven weeks previously, it is usually not necessary to clip the front leg out.

In North America all breeds are prepared similarly. In Britain, however, it is difficult to give exact detail on show preparation for breeds other than Holsteins and Friesians. The same basic rules apply so far as washing and cleaning go, but fashions in clipping are changing fast. Not so long ago, only the tails and udders were clipped out, but now several of the

Ayrshire exhibitors and at least some of the Channel Island breeders are clipping the heads out too. Only the Dairy Shorthorn breeders have not yet got around to clipping out heads.

I can only suggest that you study what is happening at the major shows and take your guidance from that. I expect that five years hence all these breeds will have evolved a clipping system which is virtually identical to that for the black and whites. The pictures in the colour section between pages 42 and 43, which were taken at recent European Dairy Farming Events, are a useful guide as to the present state of play in Britain.

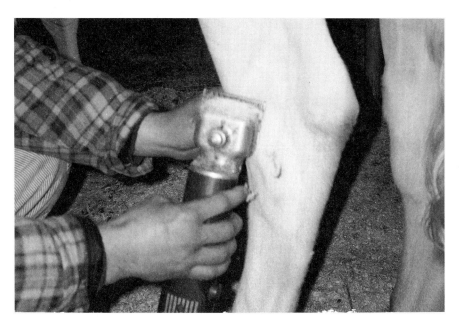

Clipping the back leg

PACKING YOUR SHOW BOX
AND OTHER NECESSITIES

IN THE run-up to your show it is advisable to start to collect all the paraphernalia which will be required. First and foremost you will need a wooden show box. If you haven't got one, I would recommend that you look around for a suitable secondhand one. If you can find one this way it will almost certainly be of superior quality to the modern boxes.

Sizes vary, but I would suggest that one in the region of 42″ long by 24″ wide and 24″ deep (105 cm × 60 cm × 60 cm) will be about right. It needs to be sturdily built, because as well as holding all the smaller items you will find it often doubles as a seat while at the show.

The box should have two compartments, plus a small tray on top of the smaller compartment. Into your smaller compartment will go the show halters, cloths and polish. In ours the electric kettle is an essential item, as all the major shows now have points for kettles. (It's surprising how many cups of tea and coffee are consumed at these events.) It is always advisable to take some loo paper, soap and a towel.

Into the big compartment go the dandy brush, curry comb, scraper, washing cloths, cattle shampoo, hair dryer, sheets for covering the animal and any necessary rugs. The electric clippers should be packed—usually a big pair and a small pair. You will also probably need an extension lead. A hose for washing down should also be included. Another essential is a folding chair, as there is never any shortage of bodies to sit in it.

The tray is very useful to hold the smaller pieces of equipment, as well as the veterinary aids which never come amiss.

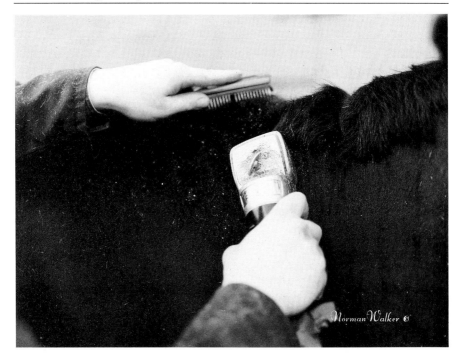

Short back and sides—a general tidy-up ahead of the big day
(Norman Walker)

Into it will go the syringes, injections, tubes of antibiotics in case of illness as well as talcum powder and hairspray. You will also need a pair of special scissors for dressing cattle plus such small items as a notebook, pen and staple-gun. Also put in anything small which would be in danger of getting lost in the larger compartments.

It is also a good idea to carry such mundane things as a hammer, a variety of nails and screws, a screwdriver, pliers, jubilee clips and adhesive tape. Fuse wire and fuses have been known to be pressed into service. Almost invariably you will find that the one thing you need is the very thing you don't have. Add it to the collection in the box before the next show. It's most unlikely that you will ever need it again, but maybe it will come in useful for someone else. The nice thing about shows is that exhibitors are very good at helping each other out, and there are very few things which cannot be produced from someone's show box. It matters not one jot that the contents of the box are tidy when you leave home, as after a couple of days at the show you will find it rather resembles the inside of your average teenager's bedroom. Among other things it becomes a very useful receptacle for dumping such items as Mother's raincoat and Auntie's shopping bag when they attend the show.

Once you have your box packed, and this can be done well ahead of time, it's time to get the rest of the necessities together. First the tools for the job. You will need a pitchfork, a manure fork, a brush and a shovel. If you can manage it, a wheelbarrow is worth its weight in gold at most shows.

For the other end of the cow I reckon three buckets for each animal is about right: that's one for water, one for dry feed and one for wet beet pulp. You will also find that a larger container is very useful for soaking the following day's ration of pulp. A plastic silage additive barrel sawed in half makes an ideal vessel for this. Most dairy farmers now use containers for things like dairy detergent as water buckets because they are nice and large, but for the concentrate and pulp feeding I prefer flat show-buckets.

Once you have decided how many cattle you are taking, you'll be able to work out what you will need in the way of fodder. In England the bulk of the show diet is usually supplied by hay, although silage is increasing in popularity as more silage clamps are open in the summer since the advent of buffer feeding. For the shows that do not involve milking trials—which need a different approach—I have always found hay to be more satisfactory, as it is so easy to handle. For ordinary shows coarse mix—several good brands are available—is used for most of the concentrate ration. If the mix is molassed, so much the better. Dairy cake can be added, depending on how much milk the animal is giving. Wet sugar beet pulp is a very important part of the diet at shows as it has a two-fold effect. Firstly, it helps keeps the bowels open, and secondly there is nothing better at filling a cow out before she goes into the ring.

We allow the following quantities per animal per day:

3.5 lbs (1.6 kg)	coarse mix per gallon (4.5 litres) for the first 4 gallons (18 litres) of milk the cow is giving
3 lbs (1.4 kg)	diary cake per gallon of milk given in excess of 4 gallons
4 lbs (1.8 kg)	dry sugar beet (to be soaked)
25 lbs (11.25 kg)	hay
	OR
50–60 lbs (22.5–27 kg)	silage

From the foregoing you should be able to calculate fairly accurately what you will need. Once you have done that, add another 30% for good measure. There is nothing worse than being short of feed towards the end of a show. You will also find that occasionally what you have with you is much more palatable to someone else's cows, and it's nice to be able to pass on some feed without running short oneself. For some reason the occasional animal will refuse point blank to eat what her owner has brought, but will almost break her neck to reach someone else's fodder. Never begrudge what you part with in this way as the probability is that at some time in the future you'll be on the receiving end of the equation.

Before you leave home you will need to know whether the show provides straw for bedding. Some do, but many still do not. If they do, of course, it makes life much simpler from a transport point of view. The requirement is something like one bale per animal per day. You will also need some short wood shavings (not the long curly kind), as these are an absolute necessity when laying down a bed for a show longer than one day.

If you have milking animals in your show team you will need a milking unit, possibly two if you have a large team. Since the advent of the parlour milking these units have tended not to be in regular use; therefore they should be carefully checked before you leave home. To arrive at the milking parlour at the show for the first milking and to find that the pulsator does not work is a cause of much frustration, yet it is still a regular sight. If you have one, a spare pulsator in the show box will never come amiss.

You will also need washing buckets and cloths, and some people carry a strip cup. The Royal Show now has a special line to spray udders after milking.

Make sure that all your equipment is marked, as it can quite easily get mislaid in the general hurly-burly. Some folks have their buckets nicely painted with their farm name; ours are simply marked with two pieces of coloured plastic tape around the handles. We use red and green, and everything, including the wheelbarrow and milking unit, has the same mark. Several other farms use this method, but with a different colour combination. This is not as permanent as paint, but the marks are simply for identification, not there to prevent theft.

By the time you get all this together you will find you have quite a pile, but you haven't finished yet, as you have your own necessities to think of. If it is simply a one-day show, then you haven't too much to concern you. You will need a boiler suit or overall of some kind (a spare one is a good idea as well), some oilskins and your wellingtons. There is also your outfit for showing, and this depends on the breed. In Britain, for instance, the British Holstein Society recommend that white shirt and trousers should be worn, Holstein Friesian Society dress is white shirt and black trousers, while the Ayrshire Society prefers white shirt and brown trousers as their official dress. The other dairy breeds at the moment are still led by attendants in white coats. Hard-soled shoes or boots are also to be preferred in the ring, not rubber boots or sandshoes.

It is advisable to take flasks of coffee and sandwiches or whatever you think you need for the day. Some bigger shows have catering facilities laid on but at best the quality of these is uncertain, while the prices tend to vary from the expensive to the daylight robbery category. If it is a major show lasting several days, many more clothes will need to be packed.

The facilities for herdsmen tend to be on the basic side, although some of the major shows have thought it worth their while to provide cubicles which, although hardly the Ritz, are a big improvement on sleeping with the cows. Some hardy souls still seem quite happy to bring their sleeping bags and to sleep in the cattle lines, but these are usually the younger exhibitors. This problem of accommodation at shows has other solutions too. Most big shows have caravan parks where it is possible to bring the whole family if so desired. If these parks are close to the cattle lines, so much the better. If you can persuade your wife to do the cooking you will find it much cheaper than eating at the establishments at the show. It's even better if you can convince her that it is her annual holiday. A few exhibitors take bed and breakfast off the show ground, but I have never found that to be an option, as work at the show usually starts in the lines around 5 am and there aren't too many to serve you breakfast before that!

If it is a major show, you are required to arrive at least a day before the show commences, and many keen exhibitors get there two days ahead to give their cattle plenty of time to settle down.

If there is a deadline for arriving at the show, allow plenty of time. If it is your first show, you'll be amazed at how long it takes to load the lorry. It is usual to load the gear in first, then the cattle. Once onto the road there seems to be a rule which says that the later you are in leaving, the more holdups you will encounter on the way.

Ray Brown (left) with his 1993 European Dairy Farming Event Champion, Bidlea Klaus Cas 2. Holstein Friesian Society President Billy Kilpatrick and the author (then President of the British Holstein Society) hold the cup (A. Bishop)

Top of the line at the National Holstein Show, 1991 (Tim Bryce)

Pat Rosser's Champion Jersey, EDFE, 1991 (Simon Tupper)

A shy Graham Madeley hides behind his Shorthorn Inspection Champion, EDFE, 1991 (Simon Tupper)

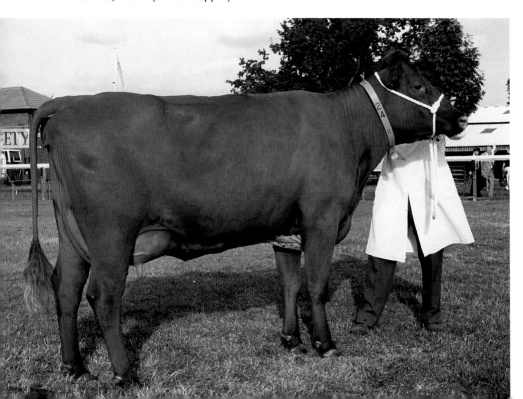

Colonel Chris Watson with his Champion Guernsey, EDFE, 1992
(Simon Tupper)

Hugh Woodburn with the Exhibitor Bred Champion at the 1991 EDFE
(Simon Tupper)

Celebrations at the 1992 Royal Show after British Holsteins had won the Burke Trophy for the best pair of dairy females

Hayleys Cynthia 22 (Bridon Astro Jet), winner of the Three Day Economic Milk Production Trial at the EDFE, 1993 (Simon Tupper)

A Champion from Canada, Browndale C.M. Reta

The Jersey team winners of the Bledisloe Trophy, EDFE, 1991 (Simon Tupper)

Faces of the future: Claire Davidson, William Armitage, Andrew Robinson and John Garnet in the Pentathlon, EDFE, 1991 (Sheila Metcalfe)

Dick Howe from Vermont casts his eye over the junior handlers at the 1992 National Calf Show (Sheila Metcalfe)

Twins Shirley and Sarah Hitchon, First & Second in their Junior Showmanship class, with James Lambert of Herdwise, who sponsored the class (Sheila Metcalfe)

Betty and Islwyn Griffiths, breeders of Cardsland Eclipse Flo, unbeaten in black and white classes in the UK in 1992 (Tim Bryce)

A sunny day at the English Royal, 1979 (Tim Bryce)

Charles Reader with his Supreme Champion, EDFE, 1991. Bob Steven, chairman of the Milk Marketing Board, presents the trophy, while Charles Stewart of Midland Bank holds the cheque (Simon Tupper)

ADVERTISING

ONE OF the benefits of showing is that it lets many people see your stock without having to travel to the farm. Indeed, some people think that showing is the cheapest form of advertising.

Jimmy Hodge (left) with Adrian, his right-hand man, displaying their wares
(Sheila Metcalfe)

Mike Carr (centre) with helpers Jenny Frampton and Martin Peach at the Scottish Winter Fair 1992 (Sheila Metcalfe)

Opinions differ as to what constitutes the ideal sign for a show, but it is usual to have a head board of some sort with at least the herd name on it. These boards vary from the simple two foot square giving the bare details to some so large that you nearly need a crane to put them up. There is at least one going around in England which requires a mechanical engineer to assemble it.

It is virtually impossible for an exhibitor to spend all his time in the cattle lines so some identification is necessary to let the public see what's what during his absences. Some larger herds also have specially printed brochures which can be left on the show box. Many of these will be picked up by the public and you never know when a contact may ensue from this. Even a supply of business cards on the box ensures that your telephone number is widely circulated. It is surprising how many enquiries at a later date arise from a show.

ARRIVING AT
THE SHOW GROUND

YOU FINALLY make it to the show ground. If it's your first show you will almost certainly feel like a fish out of water. Stay calm, relax and above all don't feel intimidated. Look around for one of the cattle stewards and ask for his advice as to where to tie up your animals. If it is a major show it is very likely that a place will have been already allocated to you. If not, you will be told where to go. This is also the time when the veterinary surgeon will examine your animals and when you will be expected to produce any documents the show may require.

Make beds up for your entries as quickly as you can. There is no need at this stage to spend too much time on this. Simply make a rough bed and unload your animals and tie them up in their allotted spaces. If they've had a long journey, they will probably be tired, so give each a drink and a little hay and leave them to lie down and rest.

Unloading the gear required usually takes much longer than actually unloading the cattle, so find out where it is to go—this will often be behind the cattle, but some shows have facilities for storing it in front. When you unload, take care to stack so that each separate lot is easily accessible. It's all too easy to pile everything up and then discover the very thing you want first is at the bottom of the pile. Don't put anything on top of the show box, as you'll very soon need something from it. Take the chair out first and when everything is unloaded and the cattle are fed, have a seat and a cup of coffee for a few minutes.

If you keep an eye on what is going on around you, you will soon discover a pattern in what at first appeared to be chaos.

45

Shiek Lorraine—(above) in her working clothes and (below), after calving, in her Sunday best (Norman Walker)

Introduce yourself to your immediate neighbours in the lines. You are going to see a lot of them. The ones a bit further away you will no doubt meet in due course. If you have arrived at the show well ahead of time, don't be in too big a hurry to disturb your entries, especially if they have had a long journey. They appreciate being allowed to rest for a while. If they, like you, are appearing at their first show, it will be quite a culture shock for them too.

If you have in-milk animals it is a good idea to let them rest until their milking time; then after milking the washing can begin. You can utilise your spare time by putting up advertising material and name cards showing all breeding details and probably milk yields. Now I know you washed and rinsed your animals nicely at home, perhaps even had them looking immaculate before they were loaded. Don't expect them to look like that when they get to the show; in fact it's quite likely that you'll hardly recognise them and you'll wonder if you ever washed them at all. So start all over again.

Using plenty of water, give a good shampooing all over and then thoroughly rinse the animal. As mentioned previously, it can't be stressed enough how important this rinsing is as any soap or shampoo left in the hair will cause flaking and dullness. Brush off the water thoroughly, and then if the weather is at all chilly, sheet your animal up, as the average cattle shed seems to have been especially designed to attract the cold in winter. In high summer the same building will be several degrees hotter than an oven. In this circumstance it will sometimes be good to take the stock outside and spray with cold water.

Once your animal has dried off nicely, every opportunity should be taken to do some work with the dandy brush. This is the most important single item in the box for putting a sheen on the coat. My old grandfather always insisted that a good grooming was as good as a feed of corn for putting a sheen on an animal, and that theory still holds good today. This is also a good time to administer any further trimming which seems necessary.

MAKING THE BEDS

IF YOUR animals are to be at the show for more than one day, it is very important to make proper beds up. This can be done at the earliest convenient time after the animals have rested.

It is not just a matter of shaking a bit of straw underneath them and hoping for the best. Making a proper bed up is a skilled job and can't be done in a few minutes. To do the best job you need plenty of dry straw and some short wood shavings. The other important thing is a manure fork.

First work out the size of the bed. It should be at least 2 feet (60 mm) longer than where the animal's back feet will come when she is tied up.

Put down a good bed of straw, then a layer of shavings and shake the two together; then beat them down level with the back of a fork. Make sure that the back of the bed is straight, as nothing looks worse than a crooked bed. Add another layer of straw, then some more shavings, then mix and beat down again. Keep doing this until your bed is built up to about 1 foot (30 mm) deep. If you have done the job properly, by the time you have finished, the bed should be very firm so that when you stand on it you can still see the soles of your shoes.

A good firm bed makes your animal look better from a spectator's point of view, and it also, being deep, allows the urine to drain away without wetting the bed. When the animal defecates, by sliding the fork underneath it should be possible to pick up the pile intact almost without disturbing the bed. If the bed is correctly made and you can't do this, you haven't got the diet quite right.

A fresh layer of straw where you have removed the dung will leave things in order for the next time. This is where your wheelbarrow comes into its own, for as each lot of manure is removed it is better to put it immediately into the barrow rather than in a pile behind the cows. When the barrow is full it is a simple matter to wheel it outside and tip it onto the inevitable pile, which will be removed at least once a day by the show organisers.

PLANNING THE TIMETABLE

B Y NOW you will have had a look at the collecting ring and been given your catalogue numbers and a schedule of judging times. Look in the catalogue and write on the back of each number the name of the animal and its particular class. You may think this is not necessary, and if you have have one or two animals perhaps it isn't, but if you have a large show team it is quite easy to get confused. This is especially the case if you have helpers who are not familiar with the cattle.

Sit down again and make yourself a list in your notebook as to approximately what time each animal ought to be in the ring. Look at the catalogue to see how many entries are in each class, as inevitably the greater the number, the longer the time it will take. One or two of the bigger shows may give a time for each class, but remember these are at best approximate and often wildly optimistic. As a rule of thumb, under a competent judge each class will take between 20 and 30 minutes depending on the numbers.

You are familiar enough with your animals to have an idea as to how much milk they need to be carrying for their udders to look at their best. This will vary quite considerably and will depend to a large degree on how long since the animal last freshened, as one recently calved ought to be giving more milk than one which has been calved six months or more and will therefore need far fewer hours between milking and entering the ring.

It is a common mistake for beginners to put too much milk on a cow, and if there is any tendency for the teats to poke this will exacerbate the situation. It is impossible to lay down any hard and fast rules as to how many hours are right, because each animal will probably vary. Once you have determined how many hours, work backwards from the time you expect to enter the ring. If, for instance, your animal needs to be carrying 12 hours milk and her class is at 2 pm you will see that the ideal time to milk her will be 2 am.

Don't be too put out about this, as by the time you have worked through a decent-sized team you may find you are milking at midnight and 5 am as well. Most hardened showmen don't expect to get to bed at all on the night before judging.

Faking udders is severely frowned upon. For the uninitiated this means such things as glueing and setting teats and using any other unnatural substances or methods to emphasise or enhance the cosmetic appearance of the mammary system.

Many show societies, especially in Britain, take a very hard line on tampering with the udder. Most do not allow any further animals from an exhibitor charged with tampering to be paraded before the judge.

It is, however, quite in order to attempt to balance an uneven udder by having different timings for each quarter. If, for instance, your animal is light in a front quarter, it is quite legal not to milk the light quarter properly out at the last milking and attempt to level things up that way. It is surprising what can be achieved by this method. I was once in charge of a breed club team and the owner of one of the animals was too busy to come to the show, but he was quite happy to send it. When it came out of the wagon I was handed a piece of paper by the wagon-driver. On it was a diagram of a cow's udder with the four quarters clearly marked. There were also instructions as to how many hours of milk needed to be in each quarter to achieve the desired effect. I was very alarmed when I saw the cow and the paper showing hours that varied from one in one quarter to eighteen hours in another. That one really took some calculating, but to be fair I was not only very surprised, but very happy to see that by judging time the cow had a beautiful udder, and looked absolutely level.

You have been looking carefully at your entries to make sure there aren't any problems and you have been keeping your animals clean. This usually involves washing every morning after milking time, as it is surprising how dirty an animal can get overnight. For the rest of the day the timely removal of the manure is usually enough to do the trick. You will often find that it is possible to have an hour or two off duty throughout the day by sharing 'picking up duties' with a neighbour, a reciprocal arrangement which allows him to have a little time off too.

Shiek Lorraine: setting up for a picture (the photo is on page 46)

IN THE RING

S HOW DAY will eventually dawn and in fact if you happen to have a large number of entries you'll find that it dawned several hours ago. Before the judging have a general wash-up and brush-up for yourself, and put your show uniform on. Having done that, put on your spare boiler suit in order to stay clean for as long as possible. It's surprising how much dirt your clothing can attract in the last few minutes. Also it is a good idea to carry a small brush and a cloth in the pocket of your show uniform, as it is remarkable how useful they may be before you get to the ring.

In Britain the British Holstein and the Holstein Friesian are increasingly more similar, and in fact anyone who has stayed with the old type British Friesian has grave difficulty in finding somewhere to show them. The methods of presenting both British Holsteins and British Friesians to the judges are now identical, as the increasing popularity of the Holstein has led to most shows now being judged by judges with Holstein leanings.

This is the time when what you have is tested against the rest and when dedicated preparation can often show a bonus. It is also the time when the leadsman plays his part on stage, so to speak.

No two animals are alike. Some approach the ideal very closely, some have minor faults, others have distinct weaknesses. The judge is trained to detect, describe and evaluate the various parts of the animal. He will assess one against the other and balance fault against fault where there are any. For anyone with a large team, once the judging has started it is a case of all systems go. Take note in the collecting ring as to whether you are to enter the ring in age order or catalogue order. Sometimes this will be the same, as some show societies catalogue according to age; others catalogue in alphabetical order of the exhibitors. Whichever method is used you'll 'go where you are put', as the saying goes here.

If there is no set order, try to get into the ring as early as possible, first if you can, although if this is your first show I am sure you will be much happier following a seasoned campaigner. Lead at a graceful walk in a clockwise direction and walk in quickly, initially leading with the right hand. Hold the animal's head just high enough for impressive style and attractive carriage. Once well into the ring, change hands on the halter and walk backwards, but not too slowly. If the judge gives you any orders, carry them out at once.

Although the cattle are being judged, never forget that the leadsman is on display too, and some practices are frowned on and discriminated against. Avoid crouching or leaning as these detract from the animal and avoid 'grandstanding'. This is the practice of continually pulling an animal out of its line towards the centre of the ring and obscuring another animal that has kept its place. It is a ploy which is akin to gamesmanship. If you are moving too slowly you may also be asked to speed up. When you are in the ring always remember the judge is in charge and he expects unquestioning obedience from the leadsmen. He will tell you when to stop and start.

You will find that the judge will usually first have a good look at all the animals from the centre of the ring. Then he will handle each animal in turn and then ask questions about age, date of calving and such like. When he inspects your animal, be sure to turn her head slightly towards him when he moves forward to feel the skin texture, as this has the effect of slackening the skin. Also step away from the animal's head when the judge moves around to the front. If your animal has a fault, do your best to minimise this by the way you set her up. Pressure on the loin with the free hand will make an animal with a high back level up for a minute or two, but try to avoid too much of this 'touching down', which can be counter-productive by continually drawing the judge's attention to the fault.

When parading, leave a six-foot gap between yourself and the animal in front if at all possible; if the class is very large this distance may have to be cut down. When lining up, always go on the right-hand side of the animals already there. Keep a two-foot gap between yourself and the next animal and keep in a nice tidy line. If you happen to be pulled in first, the ring steward will indicate where you are to stand. Once in line, set up your animal as quickly as possible.

After the judge has had a good look at the animals, he will usually draw up a short list, and if you are lucky you will be pulled into the centre of the ring where indicated by the ring steward. Change hands on the halter and walk in quickly. Most judges select six or eight animals to compare side by side, after which they usually start them off parading again in likely class order. In this parade once again change hands and walk back-wards. Once you are lined up do not relax, as you must remem-ber the judge is never finished with the class until the rosettes are given out. It is all too easy to let an animal 'fall apart' if the handler ceases to concentrate and this can sometimes lead to

Ken Empey from Canada at work on the senior heifers, National Holstein Show 1992 (Sheila Metcalfe)

the animal being demoted in the final line-up. Once the rosettes are handed out, it is usual for the judge to give reasons as to why he placed the class the way he did. When he is doing this the animals will be on their way out the ring. Once again the correct procedure is to change hands on the halter and walk forwards.

If by any chance you have a candidate in the next class, it will be necessary to have someone delegated to bring that animal up to the collecting ring, as the chances are you would not have time to travel back down the lines with the one in hand to tie her up, get another and then make it back to the ring on time. Many shows are quite sticky, I think rightly so, about animals appearing late for a class which is already parading and very likely refuse permission to enter the ring. Remember that although there are stewards, and in many cases tannoy systems, it is your responsibility for ensuring that your animals get to the ring on time.

After the individual classes have been judged, the group classes usually follow, with such classes as Produce of Dam, Breeders Herd or Get of Sire. If you have entries in these, more handlers are needed, so make sure you have them organised in good time. It is general practice to stand groups around the ring with their heads towards the rail, facing outwards. After examination by the judge the groups will normally be lined up head to tail with the winning group in front, ready to move out of the ring afterwards with the minimum of confusion. If there is a close placing, the judge may ask two or more groups to line up side by side before he makes his final decision.

THE CHAMPIONSHIPS

A T MAJOR shows where young animals are entered, there is usually more than one championship, with a champion and reserve junior and champion and reserve senior being awarded in addition to the supreme and reserve supreme championships. First and second prizewinners in each class are eligible to parade for the championships and all eligible animals should be paraded, irrespective of whether the exhibitor thinks he has a chance. A full turnout makes for a colourful spectacle, and as some spectators may not have seen the judging, this gives them a chance to see the prizewinners in the classes they missed. One other important factor is that many shows are now sponsored by banks and feed firms, and sponsors love a spectacle.

Judging the championship is the climax to the show, and if you are lucky enough to have an animal in the final parade, enjoy it. The judge, if he is proficient, will already have his potential champions picked out in his mind's eye from what he has seen in the preceding classes. The animals will normally be paraded in class order, with first followed by second in class one, then first followed by second in class two and so on. Very often three or four of the best contenders will be called into the centre of the ring for the final examination and then the judge will declare his champion by slapping it on the rump. The same procedure follows for the reserve champion. While you are in the ring it is quite possible that there may also be a prize for the Best Exhibitor-bred Animal, as well as for the Best Udder.

If at the end you have won, be modest in victory; if you have not won, congratulate the winners and remember that while the judge, like the umpire, is always right, the placings are just one man's opinion.

A happy Gerald Carter and daughter with their Great Yorkshire Show Champion 1992 (Sheila Metcalfe)

If you've managed to get this far, you will find that, at the major shows at least, your job is far from over. Even if you haven't been too successful you will still have cows to milk and keep clean. Stay with your cows; they are there as an advertisement for your herd, and many a sale has taken place after the judging.

If you've been very successful you may find that you are involved in inter-breed competitions the following day. If so, you'll need to know what time they are due to be held, and of course you'll need to work out when your cow must be milked to have her looking at her best. Then, hey presto, you probably find that the good night's rest that you were looking for following last night's lack of sleep has suddenly vanished. Such is the price of success. The judging should be fun, but it is also a place for furthering your education, so keep your eyes open and your wits about you.

Most judges nowadays will come down the lines after judging to chat with the exhibitors. This is the time to ask advice about anything you may have found puzzling and as to how you might possibly do better next time. Listen to what the judge has to say. It is not the time to start a diatribe as to why you think you ought to have been placed higher.

WHEN TO START SHOWING

IF YOU have never showed before, the question arises as to when to start. Some people never consider showing unless the animal is home bred, but there is nothing wrong with exhibiting a purchased animal provided the breeder is given credit.

One of the smaller one-day shows is the usual place for the novice showman to begin. It's very nice to win prizes at these smaller shows and you and your animals gain some valuable experience. Only a very small proportion of those who keep cattle aspire to the smaller shows, much less the major ones.

These smaller shows are great fun, but if you have never showed before they can be quite daunting. This is where you come to realise that there is one basic rule—if in doubt, ask your breed steward. He may not know the answer, but he'll find out for you. These shows are usually held out of doors, so you may get very wet. If it's any consolation to you, so will everyone else.

The entry fees are usually much smaller than those of the larger shows, but then so is the prize money. Of course, the expenses should be much smaller too, as few people travel very far to compete. It is unusual to find anyone from outside a fifty-mile radius on these occasions.

You will also find that, with the odd exception, these shows are much more chaotic. There will usually be fewer classes too and you may find your four-month-old calf having to compete against ones of up to twelve months of age. It is also quite possible to find a heifer which has calved at under two years old competing against those which are a full year older, while your second calf cow may be competing with those which have had four or five calves. On these occasions the younger animals are normally at a disadvantage.

Contrary to what one might think, the one-day shows tend to be much more demanding than the bigger ones. The night before will probably be disturbed with milking at odd times, and then there is usually an early start to get to the show ground in time to get the animals washed and ready for judging. Milking facilities for after-judging generally leave much to be desired, and there's usually a queue to leave the show ground at the end. If you had to milk before you left home in the morning and then have to again when you get back, it tends to make for a hectic day.

MAJOR SHOWS

MAJOR SHOWS may seem daunting to an inexperienced exhibitor, but they are usually very well organised and stewarded, while the facilities will be excellent and the judges top flight. You will also find that the entry fees will be higher, as will the prize money. Naturally, because of the better quality of the stock on show, it will be more difficult to win.

For sheer atmosphere a major show is in a class of its own. You'll be there at least two days before the judging, probably even before the public are admitted, so you'll have ample time to make new friends, and to ask advice for anything you aren't too sure about. You'll be surprised at how freely this advice is given. It is very nice to be successful at a minor show but a ticket in a show such as a National or the Royal carries quite a lot of prestige. You will also find that your animals are much more likely to be competing on equal terms as there is a far wider spread of classes.

The classes at a National show will probably consist of something like the following:

Class 1   Junior Heifer Calf—calves under 5 months of age
Class 2   Intermediate Heifer Calf—calves between 5 and 9
            months of age
Class 3   Senior Heifer Calf—calves between 9 and 14 months
            of age
Class 4   Junior Heifer—animals between 12 and 18 months of
            age
Class 5   In Calf Heifer—there is sometimes an age limit by
            which the animal must calve
Class 6   Dry Cow—at least 6 months in calf

Class 7	Junior Heifer in milk—having calved not older than 28 months
Class 8	Intermediate Heifer in milk—having calved not older than 32 months
Class 9	Senior Heifer in milk—having calved not older than 36 months
Class 10	Junior Cow in milk—having had two calvings
Class 11	Intermediate Cow in milk—having had three calvings
Class 12	Mature Cow in milk—having had four or more calvings
Class 13	Total Performance class—includes points for both past production and inspection on the day

You will see from the list that it is much easier for your animal to compete against those of a similar age than it is at a small show.

There will usually be specials for Best Udder in show, Best Red and White Animal, Best Animal Bred by an Exhibitor as well as Champion, Reserve Champion and Honourable Mention Awards.

END OF SEASON AWARDS

AT THE end of each show season, most countries run a competition to find the Cow of the Year. Known as the All Britain, All American, or All Canadian, this competition is based on photographs, with the show records of each cow available for the judges.

Although each country has its own rules, they vary very little. If you have a good animal it is always worth while having a professional photograph taken, preferably during the show season when the cow is at her peak. There are various categories in the competition.

The set of rules reproduced on pages 66–67 is used in selecting the All Americans and gives a rough idea of what is required. If you have a good animal, why not enter? You'll never win the raffle if you don't buy a ticket!

Shown below and on the facing page are some of the All-American nominations for four-year-olds in 1992

Miss Rosey ➤

◄ *Sequa Citamatt Tracy-et*

Paulo Bro Sir Bud Emily ➤

◄ *Boblo DLB*
Counselor Delsie

Merkley Starbuck Whitney ➤

Important—Completed Form will be required with each entry

OFFICIAL ENTRY FORM - ALL-AMERICAN AWARDS - 1992

☐ **Picture Attached**
☐ **Picture Sent By Other**
(If so, indicate from whom
and photo ID #)

Closing Date For Entries
October 31, 1992
(Except by arrangement)

Class In Which Competing

Name of Animal..

Registry No. ....**Date of Birth** ....

Sire..**(Registry No.)** ....

Dam...**(Registry No.)** ....

Name of Exhibitor ....

Co-Owner..

Address ....

Phone #..

Name of Breeder ....

Address ....

Complete Show Record for Year (please give all results regardless of placing along with production records and classification score for animals in older classes.)

Signed ....
(Exhibitor)
(See complete rules on other side)

THE 1992 ALL-AMERICAN COMPETITION (Open)

68th Annual Awards

Eligibility

(1) First through sixth place winners at the International Holstein Show.

(2) First through third place winners at the seven U.S. National Shows.

(3) First place winners at State Holstein Shows, Regional Shows and State Fairs. The foregoing also covers U.S. owned animals competing at the major Canadian shows through the Royal Winter Fair, also Canadian owned animals qualifying at U.S. shows and Canadian owned winners at their major shows who have shown in the United States.

Classes In Competition

All-American competition will consist of the following 19 classes:

Aged Cow (Born before 9/1/86)
5-Year-Old Cow (Born 9/1/86 thru 8/31/87)
4-Year-Old Cow (Born 9/1/87 thru 8/31/88)
Senior 3-Year-Old Cow (Born 9/1/88 thru 2/28/89)
Junior 3-Year-Old Cow (Born 3/1/89 thru 8/31/89)
Senior 2-Year-Old Heifer (Born 9/1/89 thru 2/29/90)
Junior 2-Year-Old Heifer (Born 3/1/90 thru 8/31/90)
*Fall Yearling Heifer (Born 9/1/90 thru 11/30/90)
Winter Yearling Heifer (Born 12/1/90 thru 2/28/91)
Spring Yearling Heifer (Born 3/1/91 thru 5/31/91)
Summer Yearling Heifer (Born 6/1/91 thru 8/31/91)
Fall Heifer Calf (Born 9/1/91 thru 11/30/91)
Winter Heifer Calf (Born 12/1/91 thru 2/28/92)
Spring Heifer Calf (Born on or after 3/1/92)
Yearling Bull (Born 9/1/90 thru 8/31/91)
Bull Calf (Born on or after 9/1/91)
Produce of Dam (2 animals, either sex)
Best Three Females bred by Exhibitor
Senior Get of Sire (4 animals, not more than 2 bulls)

Milking fall yearlings will compete in the Junior 2-Year-Old Heifer class.

(Holstein World *reserves the right to combine classes as the Nominating Committee deems necessary to ensure quality competition.*)

Entry Form Required

It is the responsibility of the owner/exhibitor to submit a completed entry form for each animal or group in order for it to be considered for All-American Nomination. Entry forms appear in *Holstein World* and are available on request from the *World* offices.

Photographs

Together with a completed entry form, owners must also submit suitable pictures of their animals qualified for entry in the All-American. Pictures must be taken in the current year and show animals in substantially the same stage as when making their show ring appearance. **Color** photos will **NOT** be accepted and there is no advantage in sending a large, unwieldy picture.

"Composite" or "head to tail" pictures of groups are not eligible for All-American competition.

Glossy prints about 5" by 7" are preferred. Duplicate pictures are requested for animals competing in both the Junior and Open All-Americans.

Pictures submitted for these awards are to become the property of *Holstein World* and cannot be returned.

Entry Deadline

Closing date for receiving entries and photos is Oct. 31, 1992. In the case of animals being exhibited at later shows, contact Lori Brown at the *World* office to make arrangements.

Procedures

A Nominating Committee will carefully consider photographs of eligible animals for which an entry form has been received. The top six entries in each class are chosen for Nomination. These Nominations are published in the **World** and submitted to the All-American Panel for voting. The panel is made up of the judges of leading shows in the United States for the current season and numbers about 25.

The panel members vote separately by mail and indicate in each class their first, second and third choice. Point values are assigned as follows: Each first place vote, 7 points; second place, 3 points; and third place, 1 point. By simple arithmetic, the awards of All-American and Reserve All-American are determined in order of total points. Every nominee receiving 16 points or more is given Honorable Mention. Third place finishers are automatically High Honorable Mention.

An entry form is required for any eligible animal or group to be considered for Nomination.

Additional copies of this form available upon request.

Please submit all pictures by October 31 and mail to

HOLSTEIN WORLD
P.O. Box 299
Sandy Creek, New York 13145
Fax: 315-387-3655

EUROPEAN DAIRY
FARMING EVENT

HELD EVERY September in Britain, the European Dairy Farming Event is, I think, unique in that it is the only show in the world where milking trials are held and the results of these trials are taken into account when awarding the breed championships and the inter-breed competitions. Of all the shows I know, the Dairy Farming Event is the one which is nearest to the heart of the true stockman. It is also the only show where no young animals or dry cows are shown. It is targeted at the practical dairy farmer, and the trade stands are only allowed if they are in this line of business.

It has its critics, who claim that the results over a short period are more a test of the stockman's ability than the animal's performance, but if the animal has no potential, the best stockman in the world is on a hiding to nothing. As the actual trials themselves cover either four or six separate milkings, it is advisable to have the cows there at least two days ahead of the start of the competitions to give them plenty of time to settle down. Milkings and weighing and testing of milk are rigorously policed by the stewards.

This show is at variance from others in that many competitors do not clip their animals until after the milking trials have taken place. These are finished by about 7 am on the Tuesday and the showing classes do not begin until the next day. Thus, the 18 hours after the trials finish are quite chaotic, as everyone wants animals washed and clipped at once.

I would hesitate to advise anyone on how to feed for the actual milking trials, as there are so many different theories on how a cow should be fed in order to achieve maximum production of milk, fat and protein, which are all taken into account in the final equation. As well as the usual feed mentioned previously, such things as cider vinegar, peanuts, treated wheat, straw and even calf milk all have their supporters. One thing is certain: not everyone gets it right all the time, not even the most experienced of exhibitors, and occasionally there is a cow with an upset stomach, which makes the space behind her very dangerous.

For the actual showing part of the event, things proceed as at any show. Many of the top trophies are only open to cattle which have competed in the milking trials. There is great competition between the various breeds for the major trophies. There are also inter-breed classes for young people in teams of four. Each member has one task to perform, these tasks being clipping, showmanship, judging and theory. There is also an area to fill with advertising and other information. Marks are given for the display and the attitude of the youngsters to visitors to the stands. This is a very good competition and is keenly contested.

See the photographs in the colour section between pages 42 and 43.

USEFUL ADDRESSES

Ayrshire Cattle Society
1 Racecourse Road
Ayr KA7 2DE
Scotland

British Holstein Society
Foley House
28 Worcester Road
Malvern
Worcestershire WR14 4QW

English Jersey Cattle Society
154 Castle Hill
Reading
Berkshire RG1 7RP

Holstein Association of Canada
Brantford
Ontario N3T 5R4
Canada

Guernsey Cattle Society
The Bury Farm
Pednore Road
Chesham
Bucks HP5 2LA

Holstein Friesian Society
Scotsbridge House
Rickmansworth
Herts WD3 3BB

The Shorthorn Cattle Society
National Agricultural Centre
Stoneleigh Park
Warkwickshire CC8 2LZ

Holstein-Friesian Association of
    America
1 Holstein Place
Brattleboro
Vermont
USA

INDEX

Farming Press Books & Videos

Below is a sample of the wide range of agricultural and veterinary books and videos published by Farming Press. For more information or for a free illustrated catalogue of all our publications please contact:

**Farming Press Books & Videos, Wharfedale Road
Ipswich IP1 4LG, United Kingdom
Telephone (0473) 241122   Fax (0473) 240501**

Stockmanship   English, Burgess, Cochran, Dunne

Gives a full account of the factors influencing the quality of stockmanship on the farm.

Calf Rearing   Thickett, Mitchell, Hallows

Covers the housed rearing of calves to twelve weeks, reflecting modern experience in a wide variety of situations.

The Herdsman's Book   Malcolm Stansfield

The stockperson's guide to the dairy enterprise.

Cattle Lameness and Hoofcare   Roger Blowey

Common foot diseases and factors responsible for lameness are discussed in detail and illustrated in full colour with photos and specially commissioned drawings. Full information on trimming.

Footcare in Cattle:
Hoof Structure and Trimming   Roger Blowey

(VHS colour video)

Roger Blowey first analyses hoof structure and horn growth using laboratory specimens and then he demonstrates trimming.

Calving the Cow and Care of the Calf
Cattle Ailments   Eddie Straiton

Two highly illustrated manuals offering practical, commonsense guidance.

A Veterinary Book for Dairy Farmers   Roger Blowey

Deals with the full range of cattle and calf ailments, with the emphasis on preventive medicine.

Farming Press Books & Videos is part of the Morgan-Grampian Farming Press group which publishes a range of farming magazines: Arable Farming, Dairy Farmer, Farming News, Pig Farming, What's New in Farming. *For a specimen copy of any of these please contact the address above.*